WITHDRAWN

About the Book

A climb up the slopes of a high mountain takes you through many different climate zones. From the mild forest zone at the bottom of the mountain to the frigid, treeless world at the top, animals and plants have found many fascinating ways to survive. While one species of willow tree at sea level grows to be 100 feet tall, another, found at very high altitudes, grows to be only one inch high!

In this book Gilda Berger takes the reader on an imaginary climb up a typical mountainside in North America and shows us what lives there. Stefen Bernath's expressive illustrations evoke the many special qualities of mountain worlds.

MOUNTAIN WORLDS
What Lives There

by Gilda Berger
illustrated by Stefen Bernath

Coward, McCann & Geoghegan, Inc.   New York

General Editor: Margaret Farrington Bartlett

Text copyright © 1978 by Gilda Berger
Illustrations copyright © 1978 by Stefen Bernath
All rights reserved. This book, or parts thereof, may not be reproduced in any form without permission in writing from the publishers. Published simultaneously in Canada by Longman Canada Limited, Toronto.

Library of Congress Cataloging in Publication Data

PRINTED IN THE UNITED STATES OF AMERICA

Berger, Gilda.
   Mountain worlds: what lives there.

   SUMMARY: Describes the different climate zones of a mountain and how plants and animals have adapted to these varying conditions.
   1. Mountain ecology—Juvenile literature.
[1. Mountain ecology. 2. Ecology] I. Bernath, Stefen. II. Title.
QH541.5.M65B47   574.5'264   77-15133
ISBN 0-698-30702-X

MOUNTAIN WORLDS
What Lives There

Imagine that you are climbing a mountain in the Colorado Rockies. On your way to the top, you feel the air get colder and colder. For every 300 feet you climb, the temperature drops about one degree Farenheit. High up in the mountain, winds blow almost all the time. Snow and ice cover the ground in many places.

Now imagine that you are on a 4,000-mile hike. You are walking north from Mexico to the northern coast of Canada near the Arctic Circle. As you head north the temperature drops lower and lower. On this trip, you must hike about 44 miles before the temperature drops one degree. The farther north you go, the more wind, snow, and ice you find.

Your climb up the high mountain takes much less time than your 4,000-mile hike. Yet each journey takes you through several climates, or climate zones. The temperature, rainfall, wind, and air pressure make up the climate of an area. Each climate zone, up the mountain or from south to north, is colder, drier, and windier than the one before.

As the climate changes, so do the kinds of trees and animals. Each type of living thing first settles in the climate zone where it is best suited to live. Then, over many thousands of years, it adapts especially to conditions there. As you pass from zone to zone the kinds of plants and animals that you see change. Species disappear, and are replaced by new kinds of plants, trees, birds, and animals.

There are many kinds of mountains. In some places they are rugged, snowy, and peaked. In other regions they are rounded and green. And there are the sandy, brownish mountains of the desert. But all mountains have climate zones that change the higher you go. Every mountaintop, and what lives there, is different from the land and life at the bottom of the mountain.

lower slopes

THE LOWER SLOPES

Suppose you are climbing up a typical mountain in North America and you start from sea level. The thick green growth of plants and trees of this first climate zone on the mountain is much like that found in nearby forests on flat land. There are vast numbers of oak, maple, hickory, beech, and other broad-leaved, or deciduous, trees. Deciduous trees drop their leaves in the fall. Then they grow new leaves in the spring. The fallen leaves decay and make the soil around these trees rich and fertile. These forests cover the mountains up to one mile or so above sea level.

Warblers, blue jays, and various kinds of songbirds nest and flit among the trees of this climate zone. Deer, squirrels, cottontail rabbits, and several kinds of mice live here, too. Most of them are plant-eating animals. They live on the nuts, berries, and seeds of the trees.

Weasels, skunks, and foxes also have their homes on the lower slopes. They are meat-eating animals. They prey on the plant eaters of the forest.

THE UPPER SLOPES

About one mile above sea level, you gradually enter a new climate zone. The weather here is much harsher. At this higher altitude the air is much thinner, because there is less oxygen in it. Thinner air absorbs less heat from the sun. The temperatures are lower. The lower air pressure causes cold winds to blow.

In this climate zone, you mostly see evergreen pine and deciduous aspen trees. As you climb higher, though, you come to spruces and firs. These are all cone-bearing, or coniferous, trees. Coniferous trees adapt to the conditions at this height by growing together in groves.

You find fewer animals and birds the higher up the mountain you go. Many animals cannot keep warm in the much lower temperatures. They find it hard to find places to hide and make their homes. Less oxygen in the air makes breathing difficult. With fewer plants and deciduous trees, the soil is poor and dry, and blows away. It cannot hold as much water.

On the upper slopes, though, you do find some porcupines, black bears, elks, bobcats, and snowshoe hares, among other animals. The blue grouse is an example of a bird that is adapted to this climate zone. It feeds on the fir needles and buds of the evergreens. This bird finds shelter from the wind and cold under the low-hanging snow-laden branches of the trees. In the spring and summer it migrates to the lower slopes to mate and hatch its young.

How did animals come to live on the upper slopes?

Most of them were forced up the mountains over many thousands of years by the struggle for food and safety. The lowlands became too crowded. They needed to escape attack by other animals, and by human hunters and settlers.

Aspen leaves

Fir needles and cone

THE "ZONE OF CROOKED WOOD"

As you go still higher up the mountain, you enter what is sometimes called the zone of crooked wood. Here the winters are very long and bitterly cold. The warm seasons are short. Strong winds blow all year round. The gusts take away most of the sun's warmth from the soil.

The trees that grow at these heights are short and stunted. This is because the frost kills the growing tops of the trees each year. The trunks grow thicker. The branches grow longer. But the trees do not grow any taller.

The trees grow close together, or flat against the ground, to withstand the very strong winds. The branches bend with the gusts. Eventually they become twisted into strange shapes.

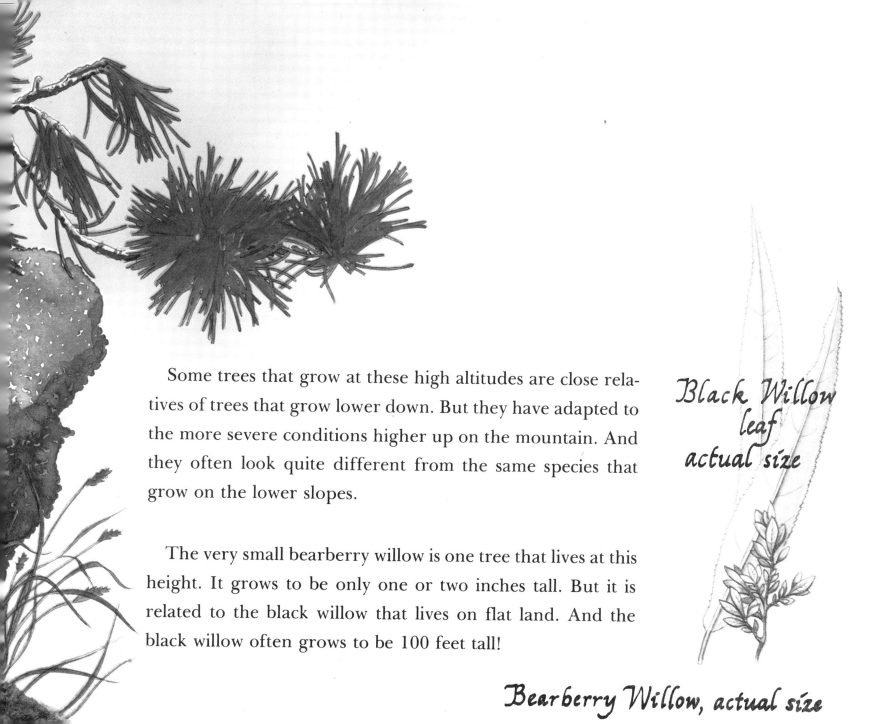

Some trees that grow at these high altitudes are close relatives of trees that grow lower down. But they have adapted to the more severe conditions higher up on the mountain. And they often look quite different from the same species that grow on the lower slopes.

The very small bearberry willow is one tree that lives at this height. It grows to be only one or two inches tall. But it is related to the black willow that lives on flat land. And the black willow often grows to be 100 feet tall!

Black Willow leaf actual size

Bearberry Willow, actual size

ABOVE THE TIMBER LINE

About two miles above sea level, you come to a new climate zone. Conditions are very severe here. It is like the northernmost part of Canada. It is icy cold and strong winds blow almost all the time. The ground is frozen and covered with ice and snow during the long, harsh winter. An imaginary line, called the timberline because above it little or no timber grows, marks this zone.

A few specialized plants can survive in this climate zone. These plants are sometimes called "belly plants" because they grow very close to the ground. Since the temperatures are low and the summers are short, they grow very slowly. At times, when rain and melting snow flood the plants, they get too much water. Other times there are drought conditions, and the plants get very little water. It may take ten years for a plant to flower for the first time.

The cushion pink is one of the most widespread mountain-top plants. Close up it looks like a tiny forest. Its cushion shape helps it to avoid the wind. It also helps it to get heat from the earth.

The cushion pinks grow about two inches above the ground. But their roots extend deep under the ground. The extra-long roots help anchor them to the mountainside and prevent them from being torn away by sliding dirt or strong winds. The roots also help the plants to find underground water.

There are very few animals near the mountaintop. The most numerous are the mountain voles, relatives of the field mouse. Other animals include various kinds of goats, sheep, rabbits, and elks. Birds such as eagles and vultures live in crags in the rocks. Another bird, the North American black swift, feeds on insects that are blown up from the lower slopes.

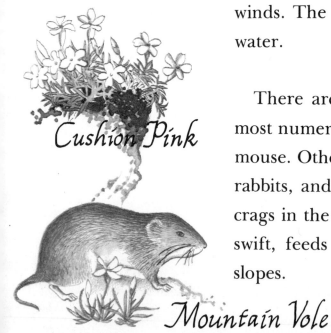
Cushion Pink

Mountain Vole

Vultures

Pikas

SPECIAL ADAPTATIONS

Pikas are small, short-eared relatives of rabbits that have adapted to life on the heights. In the late summer they snip off grasses and flowers with their sharp front teeth. They spread the plants on stones in the sun to dry. Then they gather them together and store them among piles of rocks. For the rest of the year, pikas live off their stacks of dried plants.

Large animals, such as moose, have other adaptations for mountain life. Their long, stilt-like legs are useful for walking through deep snow. Their size and heavy coat of fur helps them to withstand extreme cold. Large animals lose their body heat much more slowly than smaller animals.

Mountain Goats

Rocky Mountain goats spend all their lives above the timberline. They live on grasses and twigs. Their narrow bodies allow them to walk along thin ledges. The soft heels of their feet stick to the steep, rocky surface. They can also move over ice without slipping or falling.

Camouflage, a natural disguise, helps some animals to survive at the mountaintop. The snowshoe hare and the white-tailed ptarmigan, for instance, are brownish in the summer. That is the color of their homes of earth and stone. In winter, though, they grow new coats of white. This protects them by making them invisible against the snow.

Snowshoe Hare

White-tailed Ptarmigan

Many small animals hibernate during the winter months. The meadow vole and the marmot, a relative of the woodchuck, hibernate every winter. They find shelter under the ground or the snow. Their body temperature drops to just above freezing. Their heartbeat slows down from about 200 beats per minute to only 20 per minute. Hibernating animals conserve their body heat. They use up their stored energy very slowly. When spring comes, the animals wake up and look for fresh supplies of food.

From the leafy forests of the lower slopes to the cone forests of the upper slopes, to the "crooked wood" around the timberline, and beyond to the top of the mountain, are special places on the mountain where plants and animals live. They can also be special places for you, if you love the mountains and what lives there.

FOR FURTHER READING

Brooks, Maurice. *Life of the Mountains.* New York: McGraw-Hill, 1968.

Milne, Lorus Johnson. *The Mountain.* New York: Time-Life Books. Revised Edition, 1970.

Parnall, Peter. *The Mountain.* New York: Doubleday, 1971.

Smith, Frances C. *The First Book of Mountains.* New York: Franklin Watts, 1964.

Waller, Leslie. *Mountains.* New York: Grosset and Dunlap, 1969.

INDEX

Page numbers in italics refer to illustrations

Animals, 9, 10, 12, 13, 15, 22, 25, 28
Aspens, 12, *14–15*

"Belly plants," 20
Birds, 9, 12, 15, 22
Black bears, 15
Blue grouse, *14*, 15
Blue jays, 10
Bobcats, *14*, 15

Camouflage, 27
Climate zone, 9, 10, 12, 20
Cushion pinks, 22

Deer, 10

Eagles, 22
Elks, 15, 22

Firs, 12, *15*
Foxes, 12

Goats, 22, *26*, 27

Hibernation, 28, *29*

Ice, 7, 9, 20

Lower slopes, *8*, 10–12, 15, 22

Marmots, 28
Mice, 10
Moose, 25

Pikas, *24*, 25
Pines, 12
Plants, 9, 20, 22, 25
Porcupines, 15

Rabbits, 10, 22, 25,

Sheep, *21*, 22
Skunks, 12
Snow, 7, 9, 20, 27, 28
Snowshoe hares, 15, 27, *27*
Spruces, 12
Squirrels, 10
Swifts, 22

Timberline, *9*, 20, 27
Trees, 9, 10, 12, 16, 19

Upper slopes, *9*, 12–15, *13, 14*

Voles, 22, *22*, 28
Vultures, 22, *23*

Warblers, 10
Weasels, 12
White-tailed ptarmigans, 27, *27*
Willows, 19, *19*
Wind, 7, 9, 12, 16, 20, 22
Woodchucks, 28

"Zone of Crooked Wood," 16, *17*

About the Author

GILDA BERGER grew up in New York City, where she attended City College and earned degrees in Education and Special Education. She has taught retarded and emotionally disturbed children and has developed reading materials for her classes.

With her husband, Melvin, she co-authored *Fitting In: Animals in Their Habitats*. Her most recent book is *The Coral Reef,* in the popular *What Lives There* series. The Bergers live on Long Island with their two teenage daughters.

About the Artist

STEFEN BERNATH lives in New York City, where he is involved with many kinds of art projects, among them a popular series of coloring books about plants. A graduate of Cooper Union, he has also worked in architectural offices and as a designer of record albums. MOUNTAIN WORLDS is his first book for children.